4/₆

2.50

MW00423093

KITTY
The Raccoon

by Jamie Stamper

Illustrated by Susan Heinonen

A Storytellers Ink Book
Storytellers Ink, Seattle, Washington

Copyright Jamie Stamper 1986
ISBN 0 962 3072-0-3
Library of Congress Catalog No. 89-91699

Printed in the United States of America
for the Publisher by
GraphiColor, Inc.
Seattle, Washington

KITTY
The Raccoon

I called her Kitty.

She was tiny, all but blind and tailless. She was curious and mischievous and — eventually — the cause of my leaving home.

But she was my pet and I loved her. Something else about "Kitty." She wasn't a cat. My "Kitty" was a raccoon.

She wasn't always called Kitty, that came later. In the beginning I called her Rackety.

"Son, you have to do something about that raccoon. The animal is full-grown and can't live in the house anymore. You'll have to make other arrangements." My dad had just come home to find the kitchen floor littered with the contents of a five-pound bag of sugar. My pet raccoon had been exploring the cupboard, sampling the groceries that appealed to her senses. Rackety obviously had a sweet tooth.

With her hand-like paws, she also had a knack for opening doors and drawers, and her natural skill at climbing put virtually everything not under lock and key within her reach. "Raccoon-proofing" the house, my parents decided, could best be accomplished by dealing with the problem directly. I loved my home. I loved my mother and dad. But I couldn't let Rackety go it alone.

We went out to encounter the world together.

It had all started some months earlier when my sisters discovered a litter of new-born raccoons under an old kayak, lying upside down near a big fir tree about twenty feet from the house. Living in the Pacific Northwest, I was familiar with the squirrels, moles, raccoons and occasional coyote who frequented the woods around our home. But in my 19 years I had never adopted a wild creature as a pet. Chances are, I wouldn't have befriended Rackety either if her mother hadn't abandoned her. I also learned first-hand something about "survival of the fittest," the Darwinian theory we had studied at school.

The litter of newborn raccoons

Rackety, it turned out, couldn't see. I didn't know that at first, but the mother raccoon did, moving the rest of her litter away from the kayak. Suddenly alone, Rackety became alarmed. The cooing baby sounds that I thought of as a pleasant chirping became a raucous screeching, as the tiny raccoon searched frantically for the "family" security she had known.

Cries of animals in distress can lure predators as well as rescuers. Fortunately, the raccoon nest was close to the house. I had been checking it every day. When I realized mother and cubs had moved away, leaving one forlorn infant to fend for herself, I claimed the foundling for my own and took her home.

Raucous screeching Rackety

With Dad away on business, Mother was the target of my negotiations to keep Rackety. There was no need to worry, she was as intrigued as I with this pulsing ball of fur and quickly we had the tiny creature nestled in a small cardboard box. We then turned our attention to Rackety's survival. She obviously was hungry and near starvation and we needed expert help fast. We called the neighborhood veterinarian who advised feeding the raccoon baby formula with an eyedropper.

At the tender age of a few days, she was not bigger than a cupcake. Small snout, eye mask, bush tail neatly ringed — she was a furry miniature of all the raccoons I had seen in nature books and on television. There was one difference. This one thought I was her mother.

Our baby in her box

I had named the little refugee Rackety because the name seemed to fit. Hungry, she set up a compelling racket. Playful, she chattered happily in a chirpy, rat-ta-tat fashion. Not knowing how to converse with a raccoon, I resorted to baby talk and smiling. I smiled a lot. I'm sure, behind her mask, Rackety was smiling too.

In a few weeks Rackety had graduated from the eyedropper to one of the girl's doll bottles, then to a Playtex baby bottle, which she held quite adeptly herself — draining every ounce of baby formula. In a few months she reached adult size, dainty for a raccoon but healthy and alert.

About midway between Rackety's babyhood and maturity, I had taken her to our Vet who confirmed what I had begun to suspect. A glaze in her eyes — the white pupils — uncertain moves she made even beyond her "toddler" days, caused me to think that Rackety was blind, or close to it. "Cataracts," the Vet said. "We could put her to sleep," he suggested gently.

That, of course, was unthinkable. Now I understood why the mother had abandoned the cub. Unable to see, the animal was doomed from birth. The law of the wild seemed to me harsh and unfeeling. How, I wondered, could animals be so richly endowed with the maternal instinct and at the same time treat their offspring so callously. It didn't occur to me that animals had all they could do to protect and rear those born free of defects. Nevertheless, I shuddered at our Vet's first suggestion, but caught a glimmer of hope in his second. There is a special Eye Doctor for animals, he mentioned.

A glaze in her eyes

We took Rackety there, to a modern clinic
complete with examining rooms and operating
facilities. Yes, Rackety had cataracts. Yes, an
operation was possible — it cost $250. Would it
restore Rackety's sight? Maybe, but there were no
guarantees. In the end, I decided against it. I didn't
have the $250 but I also feared that surgery might
destroy what little sight the animal seemed to have.
Besides, Rackety managed to get around amazingly
well. You couldn't tell she was blind from the way
she behaved. She seldom bumped into anything. She
had keen senses. And she climbed trees like, well, like
a raccoon.

Rackety heading down the tree
"like a raccoon."

Rackety and I had fun during those early days of Summer. She loved to explore new places. She relished life; every day was a new adventure. I basked in her curiosity, and she seemed to sense my own excitement as we roamed the woods together. Psychologists would probably have a term to explain the bond between us. To me, we were simply a team, sharing the experiences of friendship.

By now Rackety had discarded the baby bottle and was eating dog food from a dish. Then I got a Summer job and was away during the day. At first that was all right. Rackety would greet me warmly when I came home. I'd feed her and we would play.

e roamed in the woods together.

The few times I couldn't find her right away, she would be curled up snoozing in some out of the way spot. We had a basement room where my Mother stored extra dishes and various bric-a-brac. Apparently Rackety had holed up in the storeroom to rest and someone had shut the door, trapping her inside. We don't know how long she had been locked up, but the devastation was a sight to behold. In her panic to get out she had created a collage of broken crockery on the floor. My brother, David, investigating the room's strange sounds opened the door and was rewarded with a raccoon on the head. Scrambling for footing to get past him she left a series of pawprints on his face that kept him busy for a week explaining to his classmates.

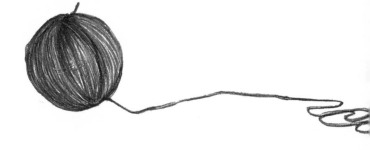

Soon, however, too much free time combined with Rackety's natural inquisitiveness led to further trouble. Things in my room began to change places. Wastebaskets were upset. Slipper traveled from closet to laundry room. Sewing boxes were pillaged. Since I usually arrived home first, I had time to repair the damages before my parents were exposed to the escalating pranks of our marauding raccoon. Then came that day my father came home early, to a kitchen floor white with sugar, catching Rackety red-handed.

My parents' solution seemed simple enough: leave the raccoon outside when I was away. I suggested an alternative: I would find a raccoon-sitter. My sister-in-law Mollie was at home weekdays. I went to see her and, appealing to her sympathy, won her agreement to keep an eye on Rackety. My relief lasted all of one day. Rushing home after work to collect my pet, I found Rackety atop the drapes in the living room, overlooking the carnage of her mischief. She was out of reach but not out of earshot of a very angry young woman. Left alone while Mollie made a quick trip to the store, Rackety obviously had made a barnstorming tour of her new surroundings. The results were all too evident: drawers opened, their contents scattered; groceries strewn wall-to-wall; house plants stripped from their containers, dirt marring the carpets. Rackety came when I called her, scampering down the drapes. Our most profuse apologies, and my offer to clean up the mess, were to no avail. Rackety was banished from Mollie's home forever.

Resigned to the inevitable, I tried trusting Rackety
to the great outdoors. Surprisingly, she adjusted well,
not straying too far from the house. One evening,
however, she didn't come when I called her. Frantic,
I combed the woods. After dinner I tried again,
calling so loudly I'm sure our distant neighbors
heard. I was heartsick, visions of Rackety lying
wounded or — worse — dead or dying, haunted me.
At midnight, when I had about given up, I heard a
scratching at the door. Rackety was cut and bleeding,
exhausted and visibly frightened, but alive. And she
had come home. I tended her wounds, fed her and
together we curled up in bed — a thankful young
man and one very weary raccoon.

Rackety restricted to the outdoors but feeling like one of the family led to some exciting interfaces with my siblings. Not as familiar with Rackety and her animal ways as I, they tended to be spooked by her pranks. She loved to ride in the car and if you were not very alert and quick when you closed the car door you would find to your surprise she was in the car with you. Getting her back out was another matter. Whether she thought it a game or a challenge, her agility and high spirits kept the action rolling until someone could find me, and I could declare her winner and reward her with a hug and a coo. One Sunday afternoon my sister Anne's breathless pronouncement that Mary was trapped with Rackety in the car and he was eating her necklace brought me up short. When I arrived, sure enough there was Rackety sitting on Mary's shoulder happily chewing away on what turned out to be a candy necklace bought at the local dime store. Mary had tried to escape, but Rackety was not easily intimidated, especially where sweets were involved.

Her outdoor hijinks weren't confined to family either. Pretty soon the paper boy started leaving the morning paper in the mailbox down at the end of the driveway. When the milkman started to follow suit with our milk and eggs, a family conference resulted.

The solution — turn half of the garage into a home for Rackety where she could safely stay when I wasn't around to keep her out of trouble. I started on the garage, penning off a section with chicken wire and making it as livable as I could. When I first put Rackety in she seemed frantic. She couldn't get out, and she was clearly unhappy. She would put her little paws through the mesh and try to open the lock. When she fixed those wistful unseeing eyes on me, I knew she was wondering why I had betrayed her. She didn't understand that the same love that united us was now the reason for her isolation.

I devised a new strategy. At night I would slip
out to the garage and smuggle her into my room. On
one occasion however, aroused by sounds from the
nearby woods, she started a staccato chatter in reply.
Upstairs I could hear doors opening and footsteps
coming. Grabbing Rackety I raced to the bathroom.
Standing in the tub hidden behind the shower curtain
clutching her to my chest we waited. Whether it was
my noisy heart or just the inevitable that gave us
away, I don't know, but when my father pulled back
the curtain and I saw the sadness in his eyes I knew
our time was coming. I tried to forestall the
inevitable by hinging one of the small panes in my
bedroom window — so that it could be swung out to
let her escape quietly in a hurry. But unfortunately, a
few days later, during a night of wakefulness, she
crawled out of my bed and wandered upstairs into
my parents' room. That was when my Dad informed
me that I would again have to make "other
arrangements."

There was no answer but for me to leave home and find a place of our own. Rackety had left her nest at an early age. I was probably overdue in leaving mine. I located a small, inexpensive house in the central district of town. I was sure the landlord wouldn't rent to me if I told him I had a raccoon for a friend, so I decided to pass her off as a cat. I renamed her Kitty. Then I was able to tell the landlord that there was just me and my little "Kitty."

Still, I worried constantly that we would be discovered and evicted. I had to leave Kitty in the house alone while I worked. But at night when I returned, she would be there and her welcome made it plain that she had missed me. Fortunately, no notes from the landlord appeared under the door.

The duffel bag on the handlebars of my bicycle was another favorite haunt. She would stay totally hidden in the bag until we stopped, then pop out and scramble to the pavement. I would then snap on her leash and we would amble off together. As we walked along, people would smile and ask if it was all right to pet her. Kitty always responded like a

lady, sitting up and waving her paws in a return
greeting. She felt people were her friends, I assume
partly because they looked like me, her "foster
mother." She would have been surprised to know
that some people feared animals and felt
uncomfortable around her.

In those days I was getting around on a motorcycle. I built a box, padded it with blankets and attached it behind the seat. Kitty quickly learned her place in the buddy-box and was soon scooting around with me as if it were the most natural thing in the world. She particularly relished the wind rushing past rippling her fur as she sat upright holding on with her front paws. We took many long rides together up into the Cascade Mountains to the East, or West by ferry to the Olympic range. She was a good companion as we shared the splendor of the outdoors together. I often wondered what she was thinking as we stopped to absorb a particularly breathtaking scene — like a snowcapped peak rising out of the rain forest above the sea? This was her world just as much as it was mine, perhaps more so. For hadn't our human species traded our native wilderness for the unnatural world of brick and mortar, concrete and asphalt? Somehow, the two of us, man and animal, peacefully sharing the outdoors together seemed right. Those were happy times.

When that first summer was over and I had left my job to return to school, I took Kitty with me — checking into the dorm with her carefully secreted in my duffel bag. For a time we got by, but my fellow students soon suspected I wasn't living alone. I mean, you don't carry dog food into a dorm room and kitty litter out unless you're keeping a pet. The guys thought I had a monkey in there. Searching for a way out, she had a habit of running her paws under the doorsill, which created a monkeylike effect to anyone passing in the hall. In due course this strange phenomenon came to the attention of the authorities and I was told to remove Kitty as it was against the rules to have pets. It seemed a bit unfair to me considering the rules that were being routinely broken in the rooms which were more questionable than harboring a raccoon.

There was nowhere for Kitty to go, so I decided to move both of us to a trailer court. Kitty lived with me there at night and spent the day in my old car parked on campus. She couldn't open the doors but did learn to roll down the windows. She would climb out of the car and play around on the sidewalk trying to make friends with the passing students. Needless to say this soon got out of hand.

I was afraid she would get run over or wander away, so I began leaving her at the trailer court while I was in class. She enjoyed exploring people's trailers and campers and would head right for their bathtubs. She enjoyed climbing up on the rim and sliding down into the tub. Being warm days the neighbors couldn't close their doors to keep her out. Someone took drastic action. That's when she lost her tail. I never found out who maliciously chopped at her beautiful tail, but it was almost completely severed, bloody and hanging by a shred of skin. I rushed her to our Vet who by now was becoming qualified in the care and treatment of raccoons. He couldn't save the tail but surgically repaired the wound and sewed a button on it to prevent further injury to the stub. Poor Kitty. Now I had a blind raccoon with no tail, and our prospects for surviving peacefully together seemed to grow increasingly dimmer.

When Summer came around again, I took a job in California. I thought renting might be easier, but the landlords I encountered wanted no dogs or cats, much less a raccoon. I protested that raccoons were the cleanest of animals (they really do wash before eating) and practiced acceptable toilet habits. By then, I had even taught my pet to use the toilet, so there was no longer any need for kitty litter.

*Kitty — now
minus her tail*

Although my brother David, who shared a bath with me during her training, still twitches when the subject comes up, Kitty's toilet training was probably no more bizarre and frustrating than that of a small child. He recalls how she enjoyed unrolling the paper and streaming it around the room, then dunking it in the water, and abandoning the wet glob in the way of his unsuspecting bare feet. And driven by a deep rooted animal instinct, she would raise a terrible fuss and scold me if after she used the toilet properly I flushed it away. Eliminating her scent that was designed to mark her territory in the wild seemed to confuse and disturb her. Despite her good habits, I could find no one willing to shelter us. Finally, in desperation, we took to living in my car, sleeping on the beach at a State Park where I could shower in the morning before driving to work.

That Summer in California turned out to be our best time together. There were fruit trees adjacent to the computer plant where I was working. Kitty would play there during the day. No one bothered her. In fact, she became quite popular with many of the workers at the plant. She was gentle, loved attention and gave affection as freely as she received it. She liked going through people's pockets, examining unknown objects by turning them over in her little paws. She could startle unsuspecting gum-chewers by taking the gum out of their mouths, chewing it herself for a time before swallowing it. Kitty was a free spirit, totally guileless and unafraid.

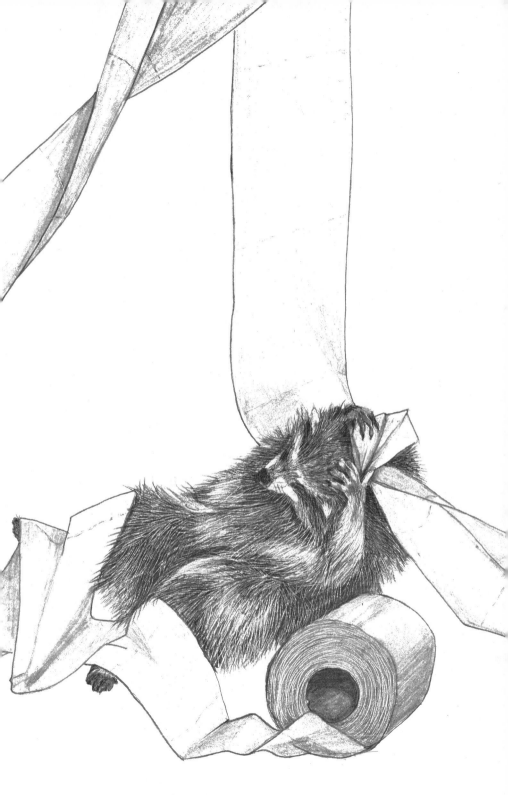

She enjoyed unrolling toilet paper.

It was this absence of fear that ultimately separated us. I recalled from seeing "Born Free" that, once domesticated, wild animals could be returned to their habitat only at considerable peril. Not having been taught to hunt and forage or to hide and defend themselves, they became easy prey to the "trail-smart" animals. At Summer's end we went back to my parents' home for a brief vacation. Kitty not being welcome indoors, the two of us spent those warm nights under the stars, sharing a sleeping bag on the deck not far from the old kayak and the big fir tree where Kitty was born. Some instinct seemed to beckon her. Being nocturnal she would leave the sleeping bag from time to time in the middle of the night, always returning in the mornings sometimes bedraggled or bearing the marks of battle. One morning she didn't return. I wasn't worried — she had always come back — but she was still missing that evening, and my alarm quickened. I spent the next 2 days combing the woods — calling and calling — even my parents joined in the search (probably with mixed emotions). Each evening when she didn't return, my hope diminished.

I brooded for days. She had slept in my bed, eaten the food I had prepared, shared my life for nearly two years. For the first time I experienced the burden of loss, the sharp incision of separation. My thoughts, I discovered, were mixed. I feared Kitty was dead and could never return to me. Ambivalently, I wondered if her instinct for life in the wild was stronger than her attachment to me. I resented even the notion that after all I had done for her, Kitty would desert me. Had she fallen victim to some predator, or had I in some way failed her? Perhaps it was my mistake, bringing her back to her native territory.

I never saw Kitty the raccoon again. But in her own way, she left her legacy. Several months after Kitty vanished, my mother called me in California. She had been driving the road that winds along the ravine near our home. She braked suddenly to avoid hitting an animal ambling across the pavement. Looking closer, she saw a small, tailless raccoon scurrying into the brush. When she called out to it, the animal paused, looked back, then disappeared into the wood.

I thought of Kitty often in the ensuing months, wondering where she was and if I would ever see her again. Gradually I realized that if Kitty had left voluntarily, she had made the right choice. No matter how close we were, man and animal — and no matter how much we meant to each other — Kitty's place was with her own kind. She would have had to leave her real mother, too, to begin a family — to fulfill her destiny. For a brief, fleeting time she had tasted a world few animals know, but in the end she belonged in the wild.

The following year I returned home. One evening we heard noises at the back door. Opening it we found three young raccoons apparently looking for handouts. They all had tails, and behind their masks were clear-eyed and inquisitive. As we fed them, I wondered: could these be Kitty's cubs? Had she sent them to me? I would never know for sure. But I like to think that somehow Kitty was saying that with my help she had made it in her own environment.

ABOUT
THE AUTHOR

JAMIE STAMPER leaves the glamorous, high tech world of computing in the aerospace industry to tell us this funny, but tender, true story of his life with a handicapped raccoon. His love for "all things living" shines through as together they challenge both conventional society and the "law-of-the-jungle" and win. We're all a little richer — that this gifted writer made the switch.